Anonymous

Report of the American Humane Association on Vivisection

and Dissection in Schools

Vol. 2

Anonymous

Report of the American Humane Association on Vivisection and Dissection in Schools
Vol. 2

ISBN/EAN: 9783337254698

Printed in Europe, USA, Canada, Australia, Japan

Cover: Foto ©berggeist007 / pixelio.de

More available books at **www.hansebooks.com**

THE
American Humane Association.

SOCIETIES OF THE UNITED STATES AND CANADA

ORGANIZED FOR THE

Prevention of Cruelty to Animals and Children.

JOHN G. SHORTALL, PRESIDENT,

560 Wabash Avenue, Chicago, Ill.

DEAR SIR:

Your attention may already have been called to the more or less public discussion concerning the effect upon our youth of those methods of instruction, obtaining to some extent in our public schools (and we fear being urged to yet wider prevalence,) whereby the facts of physiology are set forth by means of actual experimentation upon living animals, etherized for that purpose. Animals, such as frogs, pigeons, dogs and particularly cats, are dissected before mixed classes of boys and girls,—sometimes the teacher operating, and sometimes the pupils. The American Humane Association, having had its attention very forcibly called to this matter, *and realizing that public opinion must, at least encourage or discourage such methods of instruction in our schools,* earnestly desires to obtain the opinion of those who largely shape and guide the public thought. Will you therefore be kind enough to give us your judgment upon the following questions:

1st. Will experiments involving either the infliction of pain or death upon helpless creatures tend to culti- vate or to blunt the natural sensibilities of children assisting thereat?

2nd. Do you think it advisable to give to children a belief in their irresponsible power over the lower forms of life?

3rd. Do you consider it in accord with the best interests of education that children be familiarized with the infliction upon animals of mortal wounds, with the sight of blood, or the process of dying?

4th. In the teaching of children in public schools of those rudimentary truths of physiology and hygiene which pertain to the care and preservation of health, could not everything needful be clearly taught by the use of illustrations and manikins, without resort to ex- periments on living creatures?

5th. If before advanced students it be sometimes deemed advisable to expose the vital organs of ani- mals already killed, would it not seem far preferable that such demonstrations be upon animals used for food, rather than upon those whose whole existence is associated with human companionship and affection?

The American Humane Association is of the opinion that not only vivisection and the killing of animals by and before children of public school age, but also their dissection, not only neutralizes much of the work its Constituent Societies have so long been laboring to accomplish, but that such practices must inevitably operate to the moral injury of the young, and the dulling of all those finer feelings so essential to the noblest types of manhood and womanhood.

Believing that in view of the interest at stake, you will be willing to give this Association the benefit of your judgment, I am, sir,

Respectfully yours,

JOHN G. SHORTALL, *President.*

REV. FRANCIS H. ROWLEY,
ALBERT LEFFINGWELL, M. D.,
Special Committee.

REPORT.

The committee to whom was intrusted the duty of receiving replies to the circular of the American Humane Association regarding dissection and vivisection in public schools beg leave to submit the following report:

Two letters of inquiry have been issued at an interval of several months. They were identical in effect except as regards the wording of the fifth question. For the sake of accuracy it is deemed best to call attention to this difference; although it seems exceedingly improbable that the general character of replies to the first circular, as a whole, would have been essentially different had the second form of interrogation only been used.

This fifth interrogation of the first circular was as follows :

"If before advanced students, it be sometimes deemed judicious to expose the vital organs *or vital phenomena* of creatures under anæsthetics, and *killed while unconscious*, would it not seem far preferable that these be upon animals *used for food*, than upon those whose whole existence is associated with human companionship and affection?"

This implies the use of painless *vivisection* before advanced students ; and on further consideration that subject was deemed somewhat aside from the real purpose of these inquiries. In the second edition the question was changed so as to refer simply to the *dissection of dead animals* and the preferable study of the lungs and heart of a sheep or an ox, in place of animals generally used for pets, and specially put to death for purposes of dissection. It ran thus:

"If before advanced students it be sometimes deemed advisable to expose the vital organs of animals *already killed*, would it nor seem far preferable that such demonstrations be upon animals used for food,—rather than upon those whose whole existence is associated with human companionship and affection?"

Fully half of the replies received were made in monosyllables directly upon the margin of the circular itself. Others were accompanied by letters,—sometimes making slight distinctions (particularly as to the age of the pupils when dissection might be allowed,) but expressing to a greater or less degree, agreement with the sentiment prevalent in the Association.

In several instances the writers were very careful to disclaim any antipathy toward scientific vivisection in Medical Colleges, while strongly condemning its employment in public and private schools. A few others, regarding the questions of the committee as an attack upon all vivisection, have presented the Association with arguments for its defense as a method of professional instruction. If their replies are not herein quoted, it is because the writers apparently misapprehended the subject of the present inquiry—the use of dissection and vivisection in public and private schools.

Answers were received to the first circular from the following persons :

PRESIDENT DAVID H. COCHRAN, Ph. D., LL. D., *Polytechnic Institute, Brooklyn, N. Y. :*

" You are personally aware of my position in regard to vivisection for illustration. It has been forbidden in the Polytechnic Institute for many years, and no animal is permitted to be killed on the premises for illustrative purposes."

PRESIDENT M. W. STRYKER, D.D., LL. D., *Hamilton College, N. Y. :*

" While disclaiming the slightest ability to express a technical opinion, I will say from an ethical point of view that I do certainly sympathize with the considerations urged in your circular. I feel deeply that vivisection should be reduced to instances of absolute necessity, and that much of it, as practiced in the presence of those whom it teaches to neglect the rights of animals, is inevitably brutalizing."

PROF. THOMAS M. COOLEY, LL. D., *University of Michigan :*

" The whole business of vivisection of animals ought in my opinion to be brought to an end, except where it can be conducted under the supervision of experienced surgeons."

PRESIDENT E. BENJ. ANDREWS, D. D., LL. D., *Brown University, Providence, R. I. :*

" The subject is a delicate one. All experiments and operations in this department *here*, are guarded in the most careful manner ; no pain is permitted to be suffered by any creature, and no dissection goes on to which anyone could object."

PROF. SELIM PEABODY, *Late President University of Illinois :*

" In my opinion vivisection should be permitted only to such persons of advanced scientific culture and training as may wish to make it an instrument of research. . . . Vivisection *should never be used merely for purposes of curiosity, or even for illustration.* There is no place for it in any school below that which has immediately or secondarily a professional character and purpose. Least of all should vivisection be conducted in the presence of children, or persons of immature age, in grammar schools, high schools, and academies."

PRESIDENT JAMES M. TAYLOR, D. D., *Vassar College* :

" I do not think such a course in any way necessary or desirable within the limits of a general education."

PRESIDENT JAMES MACALISTER, LL. D., *Drexel Institute, Philadelphia :*

" With reference to the experimentation upon living animals in connection with elementary instruction in physiology, I beg to say that, in my judgment, it is very undesirable. . . . I quite agree with you that

the manikin and other appliances available for the
purpose of illustration are sufficient for the lower
grades of instruction and that the use of dissection
must operate in blunting the moral sensibilities of the
young people."

PROF. EDMUND J. JAMES, PH. D., *University of Pennsylvania :*

"I regard such experiments as barbarous and calculated to do far more harm, from an educational point
of view, than they can possibly do good. Any dissection
of live animals for the mere purpose of instruction is,
in my opinion, not only inhuman, but highly un-
pedagogical. The only possible condition in which
vivisection can be justified, is when it is absolutely
necessary to the actual carrying out of scientific inves-
tigations. Any vivisection for mere purposes of illustration either in public schools or in medical schools
ought to be prohibited by law. I can hardly trust myself to express my feelings upon this subject."

PROF. GEO. WILTON FIELD, *Brown University, Providence, R. I. :*

"It is not advisable to kill, dissect, or vivisect any of
the red blooded animals in the presence of young
children. Manikins preferable, otherwise alcoholic
preparations."

JOHN T. PRINCE, *Board of Education, West Newton,
Mass. :*

"Vivisection has no place in our public schools, and
ought not to be practiced there. . . . As to killing

animals before children, I quite agree with the views of your Association, but I cannot agree with it in its condemnation of dissection of animals. In the upper grades of the grammar schools and in all grades of the high school a knowledge of the structure of animals should be gained and it cannot be gained by any means so well as by actual observation of the parts under the direction of a teacher."

PROF. GEO. W. ATHERTON, *Star College, Pa. :*

" It seems to me that the practice of either vivisection or dissection in the presence of children of the usual school age is not only unnecessary, in the grade and amount of instruction that can be given in the public schools, but is altogether injurious and inadmissible. Its advantages at that stage of instruction seem to me to be very slight, while the disadvantages and injurious results upon the habits of thought and feeling of the pupils seem to me so obvious that every right thinking person must revolt against it."

REV. DR. C. W. LEFFINGWELL, *Editor " The Living Church," Chicago, Ill. Founder of St. Agnes School, Knoxville, Ill. :*

" Experiments involving pain or death of animals must blunt the sensibilities of young persons present; it is not advisable to teach children that they have a right to deal with this mystery of life for such purposes.

" I think it most undesirable to familiarize the minds of the young with the sight of suffering and dying. Those whom it does not shock it will harden."

There is no need of experimenting upon living creatures for the imparting of physiological truth ; every organ can be studied and dissected apart from the body from which it is taken, and without exhibition of the living animal or even of the dead one. It is unnecessary to take any animal, as a whole, for the purpose of instruction. Each part can be dissected separately."

Rev. Dr. Lyman Abbott, *Pastor of Plymouth Church, Brooklyn. Editor of "The Outlook," New York City :*

" I am not sufficiently acquainted with the *pros* and *cons* in the matter of vivisection to express any valuable opinion upon the subject at large, but I should think it very clear that not only vivisection, but even the dissection of animals, carried on by or before children of public school age must do a great deal more harm than it can possibly do good."

B. O. Flower, *Editor of "The Arena," Boston, Mass. :*

" It is difficult to conceive of anything more injurious to the child than allowing it to witness or engage in experiments involving the infliction of pain or death upon helpless animals. It is bound to blunt the finer sensibilities and call out the savage in the child.

" I am as unqualifiedly opposed to the familiarizing children with the infliction of pain or mortal wounds on animals as I am opposed to giving children military instruction in our schools. The child that becomes familiar with torturing dumb animals and the child who is familiarized with war, during the plastic years when his character is being formed, will necessarily be brutalized to a very great extent. I do not believe in vivisection. I believe that all experiments necessary have already been made.

"Certainly there is no excuse whatever for using aught in the public schools beyond illustrations, manikins, etc."

HON. ARBA N. WATERMAN, *Judge of Illinois Appellate Court, Chicago :*

"Civilization in its moral aspect consists in a heightened sympathy with, and consideration for, those men or animals in our power. It is impossible to train a child to indifference as regards the suffering of a helpless dog, and at the same time to be mindful of the rights of little children."

REV. O. B. FROTHINGHAM, *Boston, Mass.:*

"I have no hesitation in expressing my hearty approval of all the ideas contained in the circular you so kindly send me. Young people can get all the physiological instruction they need, and more, without hurting a single creature. It is a shame that they should be demoralized by experiments that inflict pain on the lower animals; that they should regard these animals as victims of irresponsible power; that they should early be familiarized with blood, torture or death. If vivisection is necessary, a matter that I am not quite sure about, it should be confined to skilful physicians experimenting in laboratories, or lecturing to adults, and under conditions which insure the utmost benefit with the least possible torment. If the human advantage is merely probable and the agony considerable, the advantage should be forborne. We have learned to wait for knowledge, while as to bearing pain, man, with his vast mental and moral resources, ought to be ashamed to confess inferiority to the dumb beasts."

Rev. Dr. H. W. Thomas, *Chicago, Ill.:*

" The practice of vivisection in the higher schools our country, medical and other colleges, has been carried, to say the least, to the borders of abuse, and its introduction to the public schools should be discouraged and condemned by all who have the highest good of the rising generations at heart. It is not necessary for practical instruction in physiology, and if such lessons are needed, they should be taught from the forms of life taken for use as food.

"In all young minds and hearts should be cultivated a sacred reverence for life and the kindest feelings for every creature capable of suffering pain.

"We should certainly hope that the humane sentiments of our age will create a public feeling so strong as to discourage and prevent every form of cruelty and the shedding of blood in our public schools."

Rev. A. W. Stevens, *Cambridge, Mass.:*

" No person, old or young, should inflict either wounds or death on any animal except in cases of clear necessity."

Rev. N. Seaver, Jr., *Millbury, Mass.:*

. . . "Stating the case broadly, I think that in- formation calculated to make children less humane is not profitable or even pardonable. Better more attention to the rudiments of education and fewer of the fads that are instructive and helpful, if at all, to but a *very small* minority of mature and scholarly minds. To permit children to witness what they must regard as torture is positively demoralizing. To fill their heads with scientific facts for which not one in a thousand will ever have a use, and then permit them to graduate from High School, spelling 'separate' with one 'a,' is a piece with much other prevalent nonsense."

Rev. Dr. R. A. White, *Chicago, Ill.:*

"I fully and heartily concur in your efforts to stop the practice of vivisection or dissection of animals of any kind in the public schools. Vivisection under proper restrictions may be of sufficient value to medi- cal science when performed by medical experts to counterbalance its cruelties. But anything of the kind before public school pupils, not one in a thou- sand of whom will ever study or practice medicine, is absolutely unnecessary and without reason. It bru- talizes the children, subordinates in their estimation the rights of animals to life and reasonable human care, and is of no practical value beyond what could as well be obtained in other ways. Set me down as one who loves my fellow animals, and deprecates any unnecessary infliction of pain upon them, no less than that loss of fine feeling which inevitably follows on the part of children systematically trained to hold the sufferings of animals lightly."

Rev. W. C. Gannett, *Rochester, N. Y. :*

" If the custom of vivisection is entering our public schools, I rejoice that you are taking up the matter in this way. . . . Were a child of mine attending a private school where this practice was followed, I should feel that all good it might get in other ways would be largely offset by this cruelty, and should take the child away."

Rev. Dr. Leslie W. Sprague, *San Francisco, Cal. :*

" To inflict pain may not be the result of cruelty; but it causes either deadened sensibilities, or a delight in seeing pain."

Rev. A. J. Chapin, D. D., *Omaha, Neb. :*

" I believe the business of dissection, and especially of vivisection as practiced in the public schools of all grades, to be wholly unnecessary and wrong, and am glad to use any influence which I may possess against the demoralizing practice."

Rev. J. E. C. Welldon, D. D., *Head Master Harrow School, England. :*

" I should say such experiments will undoubtedly blunt the sensibilities of children. Their power is not irresponsible, and they should certainly not be taught that it is."

PROFS. LADD AND DANIELL, *Chauncey Hall School, Boston, Mass.:*

"Without expressing any opinion in regard to what may be wise for college students, we disapprove very strongly of vivisection in grammar or high schools."

MISS FLORENCE BUCK, *Cleveland, O.:*

"I am in hearty sympathy with the effort of the Humane Association in this direction. For eight years I have been a teacher of physiology in the High School, and am convinced that both there and in lower grades, all that pertains to human physiology which comes within the scope of such instruction, may be taught from manikins and from organs of animals used for food.

"I hope the Society will also protest against the experiment so frequently described in works on Physics, that of introducing a mouse under the receiver of an air-pump. To allow students to witness the dying struggles of a helpless creature is injurious to the finer sensibilities."

MISS LILLIAN FREEMAN CLARK, *Boston, Mass.:*

"In my opinion, the sight of suffering in animals or human beings is only harmless to the bystander when his presence is necessary or desirable for the relief of the sufferer."

PROF. J. H. ALLEN, *Cambridge, Mass.:*

"My opinion is not that of an expert in the present methods of common school instruction; but it is clear and decided on the following points:

"1. It is shocking and unpardonable that anything approaching or resembling vivisection should be permitted except in professional schools, and then only under the greatest precautions as to anæsthetics.

"2. That any form of dissection of animal tissues is probably worse than useless as a basis of instruction, except in special classes of the highest grade of public schools; and for all that can be profitably taught to the ordinary pupil, plates and models are preferable on every account, size, neatness, intelligibility and precision."

PROF. WILLIAM JAMES, M. D., *Harvard University.* :

"By such experiments I should apprehend no special effect in the way of either heightening or blunting the sensibilities of average children. There are 'psychopathic' children who might either receive a haunting shock, or an impulse to cruelty, according to the bent of their weakness, from the sight of dissection, etc.

"To the third question I reply 'no,' so far as children below 16 or 17 are concerned. After that age the answer depends on the special circumstances. To *have seen* mortal wounds and death is often a vitally important experience. To be 'familiarized' with them may be unfortunate. To be familiarized with *blood*, in the case of those whom it makes faint, means the overcoming of a most deleterious weakness. . . .

"At the high school age of 17 or 18, the sight of a dead animal dissected is for *almost all boys* a highly desirable experience, ministering to a most legitimate intellectual need. With a serious teacher, I see no

possible harm, except to '*psychopathic*' subjects. I
believe vivisection of any sort to be quite out of place
in schools of *any* grade.

·· To college classes, vivisectional demonstration of the
spinal reflexes on a decapitated frog, and exhibition of
a frog and a pigeon painlessly deprived of their cere-
bral hemispheres, are invaluable. Other vivisections
(a frog's nerve muscle preparation can hardly be called
a vivisection) seem to me best .omitted.

" I believe that there goes on in medical schools a
lot of purely wanton vivisection for purposes of ' demonstration,' which the class does not see, and which
is wasteful of life and condemnable.

" I believe in keeping up a sore state of public opinion as to this latter sort of cruelty. . . . What is
needed is a great public sense of the *responsibility* of
our power of life and death over lower creatures. For
this result as much as anything depends, it seems to
me, on the example of the teacher's spirit."

PROF. H. E. SUMMERS, *Professor of Physiology, University of Illinois:*

"Children show little *natural* sensibility to pain
inflicted on lower animals ; what they have can only be
imparted to them by careful training. Hence the witnessing of the infliction of pain is decidedly harmful,
as tending to prevent their acquisition of a proper
degree of sensibility. . . . A fly may be killed cruelly, a pet dog humanely. It is impossible, even if
desirable, to prevent most children killing at least
insects ; they should therefore be taught to do it

2

humanely with full regard to the feelings of the smallest living creature.

"Children should certainly not be given a belief in their *irresponsible* power; of the power controlled by a great responsibility, yes. They will learn that they have the power despite us; and the knowledge of responsibility should come *with* the knowledge of the power."

EDWIN D. MEAD, *Editor " New England Magazine :"*

"In reply to your circular concerning dissection and experimentation upon animals, in connection with the teaching of physiology in the schools, I would say that all such work should be done with great care and under the most scientific supervision. I cannot conceive of conditions which would ever make it necessary or useful in the lower grades of any schools."

PROF. BÄR, *University of Göttingen, Germany:*

. . . "I agree fully with the American Humane Association in the opinion that not only vivisection, but even dissection of animals, killed by and before children of public school age, will inevitably operate to the moral injury of the young."

EXTRACTS FROM REPLIES TO THE SECOND CIRCULAR.

His Eminence, Cardinal Gibbons, *Baltimore, Md.:*

"In reply to questions addressed to me in the name of the American Humane Association, I beg to say that I am inclined to think such experiments as you mention tend to blunt the natural sensibilities of children assisting thereat.

"The best interests of children, in my judgment, require that they be not familiarized with the sight of blood or death, inhumanly inflicted.

"I am inclined to think that sufficient instruction could be imparted by the use of illustrations and manikins. I think it advisable to give children the knowledge as Scripture does, of the God-given power of man over the lower forms of life ; but they should be warned that this power is not absolute, arbitrary or cruel."

Rt. Rev. Bishop Alfred Barry, *Chaplain to Her Majesty, Windsor Castle, England:*

"I take it for granted that in the experiments referred to effective anæsthetics are used, and therefore that no cruel infliction of pain takes place.

"But even in that case, I should think it most undesirable to perform experiments on living animals before children just at the age at which experience proves that there is the greatest temptation to recklessness and cruelty. Even under anæsthetics there is often, as we know in human subjects, the appearance of struggle and suffering, which should be in itself offensive to a sensitive temper, and the desire of imitation, so characteristic of growing boys, is not unlikely to lead to the repetition of the experiments without anæsthetics. I cannot believe that for such physiological and hygienic teaching as is suitable to children, vivisection can be necessary. There are many things not wrong in themselves which we should keep from children to whom " *maxima debetur reverentia* ," and vivisection is one of them."

Rt. Rev. Geo. F. Seymour, LL. D., *Bishop of Springfield, Ill. :*

" In response to your inquiries, I would say, without any qualification, that the American Humane Association is right in the position which it takes as expressed by your questions.

" To reverse the policy of taking thought and sympathy for others, and particularly for creatures which cannot protect themselves and have no laws to shelter them, is most pernicious, in my judgment, in its effects upon the young, and the result must be most disastrous upon character. The plea for such atrocities will not bear serious consideration. All the knowledge of the economies of life needed by the ordinary man or woman can be readily obtained from illustrated works on the subject of physiology within the reach of all."

RT. REV. FRANCIS M. WHITTLE, LL. D., *Bishop of Virginia :*

"Such experiments I think must most decidedly blunt and destroy the sensibilities of children."

RT. REV. WILLIAM ANDREW LEONARD, *Bishop of Ohio :*

"Objective lessons in pain must necessarily deaden and dull the sensibilities of boys and girls.

"Children should be taught kindness and gentleness towards God's creatures ; they should realize their responsibility to hurt nothing.

"The sight of blood and physical agony should not be allowed to children. In Connecticut, I believe, no butcher may sit as a juryman in a murder case, and the law is doubtless based on this principle.

"Undoubtedly children can be taught all that it is necessary for them to learn of physiology and hygiene by illustrated books, manikins, etc. Indeed, I am shocked to learn that vivisection is practiced in our public schools. If it be so, then our public school system needs renovation and reformation of a very vigorous character in this direction. Our schools are not halls of dissection, nor do we pay our school taxes in order to develop public education into higher education. I am sure that multitudes of right minded citizens of all degrees and all opinions will rally to your support."

RT. REV. CORTLANDT WHITEHEAD, *Bishop of Pittsburgh, Pa. :*

"I very heartily and sincerely add my protest

against such methods as those mentioned in your circular.

"'To my mind it is absurd and fanatic to make use of any such methods with children of public school age, and I have no sympathy whatever with those who would advocate them. . . . So far from being in accord with the best interests of education, I think that such instruction will ultimately be of great injury, not only to the children themselves but to society in general."

RT. REV. THOMAS A. STARKEY, *Bishop of Newark:*

"In my judgment it is of the greatest importance that all children, boys especially, be taught carefully and with painstaking, humanity to animals. It is more than important, it is vitally necessary. Children are apt to be thoughtless; boys are often so to the verge of cruelty. Any exhibition, therefore, which is deliberately prepared and with such experiments as you describe, must, in my opinion, have the effect of encouraging this native insensibility. We may easily pay too dear for knowledge, and whatever benefits may accrue in the way of added knowledge from such methods of instruction as those you refer to, is dearly purchased by the loss of so great an element in Christian character as humanity; the chivalric feeling of the strong for the helpless and weak."

RT. REV. JOHN SCARBOROUGH, *Bishop of New Jersey:*

"I am entirely opposed to vivisection, whether in schools or in medical colleges, as a barbarous and cruel thing, unnecessary and brutalizing in its tendencies, and utterly without excuse."

RT. REV. DAVID S. TUTTLE, *Bishop of Missouri :*

" I have no hesitation in saying that in my opinion the practice of vivisection, or anything approaching to it, in the infliction of pain upon the lower animal creation, as a means of education of our children in the public schools, is much to be deplored, and should be resisted by all who have at heart the good of the race and the nation."

RT. REV. A. CLEVELAND COXE, *Bishop of Western New York :*

" I am shocked even to read the inquiries contained in your circular, and I cannot but add my name under the conviction that such abuses are as horrible in view of their effect on the young as they are in view of the tortures inflicted on the brutes."

RT. REV. W. C. DOANE, LL.D., *Bishop of Albany, N. Y.:*

" I do not believe the effect upon children of witnessing experiments upon living animals can possibly be good. It must either shock their sensibilities if they are what they ought to be, or tend to encourage them in cruelty if they have that unnatural strain in them. It seems to me that physiology can be taught and ought to be taught without such experiments, but I beg leave in saying this, that I am not opposed to vivisection, when it is conducted *under the restraints of proper regulations.*"

Rt. Rev. N. S. Rulison, *Assistant Bishop Central Pennsylvania :*

" In my judgment vivisection and the killing of animals by and before children attending the public schools, and also the dissection of animals under similar circumstances are practices which cannot be really necessary and which most inevitably blunt the sensibilities and corrupt the character of the young. Practices so abhorent to the finest feelings and injurious to the best character should be suppressed by society."

Rt. Rev. Thomas Clark, *Bishop of Rhode Island :*

" I was not aware that any such atrocity existed, as the introduction of vivisection into our ordinary schools, and I think that it ought to be forbidden by law.

" If physiology cannot be taught our children by the use of manikins and illustrations, it will be well not to teach it at all. . . . I am not sure that operating on the living subject is ever justifiable."

Rt. Rev. Henry Adams Neely, *Bishop of Maine :*

" In regard to the practice of vivisection in the presence of school children, let me say in one word that I am utterly opposed to it."

Rt. Rev. John Williams, LL.D., *Bishop of Connecticut :*

" Without entering especially into particulars, I am quite ready to say that in my view, any and all vivisection and killing of animals before children of public school age, and also their dissection, cannot but be most injurious to such children and ought to be entirely discouraged."

RT. REV. CHARLES H. FOWLER, *Bishop M. E. Church, Minneapolis :*

. . . "Cruelty is a sign of barbarism. Vivisection engenders cruelty or indifference to suffering. Therefore it reverses the order of the refining forces of civilization."

REV. DR. MORGAN DIX, D. C. L., *Rector of Trinity Church, New York :*

" I was not aware until I read your article and the circular of the Association that the method of instruction to which they refer had been introduced into our schools. Yet I cannot say that I am surprised at this latest development of the exaggerated and fantastic spirit of our times.

" The system of education of the young appears to need a fundamental reform, and it is perhaps fortunate that fads of this kind should be introduced as rapidly as possible, in order that the need of such a general and rational overhauling in the interests of much abused childhood may become more thoroughly evident to the general view. . . . I am sure that vivisection should be prohibited under severe penalty, except when performed by professional men licensed to practice it for undoubtedly sufficient reasons. As to the dissection of animals before mixed classes of boys and girls as a part of the curriculum of instruction in our common schools, I fail to see any justification for it. Children need to be taught lessons of kindness and consideration for the creatures which we domesticate and of which they make pets and companions.

"It is not necessary that the average boy or girl should be made an expert in anatomy, physiology or biology. Such studies are only appropriate for those intended for the degrees in surgery and medicine. I feel certain that all that is necessary for the time can be accomplished by models and illustrations, and that there can be no need of a display of ether, knives, blood, wounds and death.

"Upon the whole, I confess to amazement at the infatuation of those, whoever they may be, who have introduced, or deem it wise to introduce, such methods into an already overloaded system of education, and I deprecate with all earnestness the mischief likely to ensue from so wide a departure from the principles and modes of sober common sense and useful teaching.

"I trust that throngh the efforts of your Society the public may be awakened to a sense of the harm and wrong done to the rising generation, and that wise counsels will prevail over these latest outbursts of a well-intentioned but as I think most mischievous pedagogy."

REV. DR. DAVID H. GREER, *St. Bartholomew's Church, New York:*

"I think it is not wise to introduce vivisection into the public schools. It is, in my opinion, for children an unnatural and pernicious method of instruction, calculated to do more harm than good."

REV. DR. HENRY VAN DYKE, *Pastor Brick Presbyterian Church, Fifth Ave., New York City:*

"I have no hesitation in expressing my opinion that the practice of vivisection in our public schools as a method of instructing boys and girls in physiology, is simply monstrous. . . .

"There is no reason in the world why our common schools should teach physiology at all, except in its most elementary form. Children do not need to have all the reasons for keeping the skin clean explained to them. They need only to be told that they must wash in order to keep well. And then, if they go dirty, a little judicious chastisement will be more effective than a hundred "lessons in physiology." To try to teach boys and girls all about their gastric juice and lymphatic glands in the common schools, is to waste the taxpayers' money and increase the mass of half-knowledge which is so much more dangerous than plain, unassuming, modest ignorance."

REV. DR. WM. N. MCVICKAR, *Philadelphia:*

"I deeply sympathize in every effort which is being made to abolish the wrong which vivisection is committing, not only on its dumb victims, but as well upon those who in any way participate in its inflictions."

REV. DR. JOHN HALL, *New York City:*

"It is not needful to enter into details; it is enough to say that I disapprove of such processes as your circular describes, and for the reasons suggested."

REV. FREDERICK R. MARVIN, M. D., *Great Barrington, Mass.:*

"Though now a minister of the Gospel, I was educated to the profession of medicine and was graduated from the college of physicians and surgeons, 'Medical Department of Columbia College, N. Y.,' in 1870.

"In the class-room I saw vivisections so unqualifiedly cruel that even now, they remain in my memory as a nightmare. I am persuaded that none of the so-called experiments upon living animals that I witnessed were of any real value to me or to my fellow students.

"I make, therefore, one inclusive answer to your five questions and say that vivisection is seldom if ever justifiable, and is never to be tolerated in a public lecture or in the presence of the young, who are almost sure to be brutalized thereby."

REV. DR. ARTHUR BROOKS, *New York:*

"I am very thoroughly in sympathy with every effort to preserve unharmed the delicacy of feeling which belongs naturally to the young, and I do not believe that any of the educational methods call for practices which would lead to an opposite result. I cannot think that any knowledge which is gained by as great a sacrifice as the loss of tenderness and pity, is valuable to the pupils in our public schools."

REV. DR. THOMAS A. NELSON, *Brooklyn, N. Y.:*

"The result of vivisection before the eyes and minds of immature school children does little more than gratify a morbid and cruel curiosity. It leaves behind a miserably small increment of knowledge to compensate for the irreparable injury to those finer instincts and sympathies which are the patent of our nobility as man, and which lift us above the level of that inferior life, so often needlessly tortured to gratify a simulated passion for knowledge."

Rev. E. E. Gordon, *Sioux City, Ia.:*

" Thirteen years experience in teaching before I became a minister and all my work with young people since that time convince me that experiments involving the infliction of pain or death upon animals do tend to blunt the natural sensibilities of children who have anything to do with them. Emphasis should be placed upon the sacredness of all life."

Rev. James O. S. Huntington, *Holy Cross House, Westminster, Md.:*

"History makes it quite clear that such experiments will tend to blunt the sensibilities. Education means not merely crowding facts into a child but making him more *humane*."

Rev. Dr. Charles H. Smith, *Buffalo, N. Y.:*

" I am very glad the Society is taking up this question with earnestness, for this is one of the subtle ways in which the evil one is now seeking to harden the children's hearts and take away that feeling of kindness and sympathy which goes far to make the man and the Christian. We have got beyond the stage of brutal cruelty, if I may so term it. There is no risk now in undertaking to protect animals against the cruelty of the heartless driver. A word is all that is necessary. We can all remember, and it was not so very long either, when to take the part of the animal exposed one to the profane and vulgar rejoinder, if not to physical danger, from inhuman beings. Now the Society's work must be to prevent this subtler form of cruelty, which is carried on under the guise of educational advantage. It seems to me that teachers who advocate experiments of this kind only desire to *interest* the children. There is a kind of morbid desire they cater to, without taking into consideration the terrible effect upon the learners."

Rev. Dr. George C. Yeisley, *Hudson, N. Y.:*

"There is no need for the torture of living animals to teach children the rudimentary truths of physiology, and even if there were, the knowledge would be dearly bought at the expense of the moral health of the pupils."

Rev. Dr. B. F. DeCosta, *St. John the Evangelist Church:*

"I agree entirely with the American Humane Association on the subject contained in this circular. The practices alluded to are brutal and demoralizing and against the best interests of humanity."

Rev. Walker Gwynne, *Summit, N. J.:*

"I had no idea that anywhere in any country, much less in this one, were public schools made the scene of such brutalizing experiments as the dissection of living—even if etherized—animals. I gladly endorse the protest of your Society against this practice."

REV. DR. ANDREW W. ARCHIBALD, *Hyde Park, Mass. :*

" My boys would be withdrawn from any school where there was opportunity offered to see living creatures dissected for the sake of illustrating facts in the structure of animal life. My whole nature revolts against such realism in educational methods."

REV. HENRY BASSETT, *Providence, R. I. :*

" I firmly believe that the exhibition of animal suffering, whether inflicted under the guise of scientific information or not, is brutalizing in its tendency and does beyond question blunt the finer sensibilities of those engaged in the practice whether they be children or those of mature years."

REV. GEO. K. HOOVER, D. D., *Chicago, Ill. :*

" The infliction of pain or death upon a helpless creature will most certainly pervert the moral nature of children. Time was when people could lead a victim to the stake and witness his agony with comparative complacency, but a great many customs that once were not only tolerated, but readily accepted, are now utterly banished from society."

REV. DR. HERRICK JOHNSON, *Chicago, Ill. :*

" The opinion of the American Humane Association is my opinion."

Pres't Ambrose C. Smith, D. D., *Parsons College, Fairfield, Iowa :*

" While not opposed to vivisection. . . . yet it ought not to be left to the caprice of every experimenter. To let every tyro torture animals in the name of science, and to exhibit such experiments to school children is, in my opinion, revolting and outrageous."

Rev. Dr. A. S. Freeman, *Haverstraw, N. Y.* (Pastor here for forty-eight years) :

" I am fully in accord with the object designed to be secured by the American Humane Association."

Rev. Frederick E. Dewhurst, *Indianapolis, Ind. :*

" Keep the scalpel out of the hands of children, and give them Wordsworth and John Burroughs to read."

Rev. Charles A. Northrup, *Norwich, Conn. :*

" I am heartily in sympathy with the object you are seeking to attain, viz. : a public opinion averse to such methods of instruction."

Prof. Felix Adler, *" Society for Ethical Culture, New York City :*

" With the spirit and purpose of the questions contained in the circular of the American Humane Association, I sympathize entirely, with one exception. The dissection of animals after death, if undertaken for the purpose of scientific study and for the attainment of knowledge not otherwise attainable, does not appear to me likely to operate to the moral injury of the young and the dulling of their finer feelings. In elementary schools it will be necessary to resort to this practice frequently, and if the teacher approaches the subject in the right spirit, I should apprehend no evil results."

W. W. STORY, *Rome, Italy:*

"I have no hesitation in saying in reply to the first three questions, distinctly 'No,' and to the last two questions to answer as decidedly, 'YES.'

"All the facts of physiology which are needful or appropriate to be learned by children, can in my opinion be sufficiently taught by means of diagrams, models, and drawings with explanations by the teacher without recourse to the dissection of dead or living animals.

"The latter course would, I think, naturally tend to blunt their sensibilities, to render them callous to suffering, and to induce them to tamper with Life, out of an excited and unhealthy curiosity without any corresponding benefit."

FREDERIC HARRISON, ESQ., *London, Eng.:*

"I am surprised and shocked to learn that there can exist schools of any kind where young boys and girls are allowed to witness dissection of living animals under any circumstances whatsoever. I will not enter on the deep problem of vivisection as a means of research, nor do I concern myself with the various modes of producing total or partial insensibility.

"Men may differ as to the lawfulness or value of

3

acutely painful experiments on living animals, when conducted by highly trained men of science in pursuit of a definite scientific problem of great utility to the human race. And it is possible to differ as to the degree and efficiency of various anæsthetics.

" But I should have thought that all persons of decent feeling and of practical experience of the young must be agreed on the depraving effect of accustoming boys and girls to see death inflicted, to witness organic operations, and to find that the ghastly incidents of the surgical and the dissecting table are part of their manuals of education.

I can imagine nothing more certain to blunt their sense of humanity, and to surround their intellectual life with degrading association.

" Those who are parents or moral teachers know how difficult it is to extirpate the love of cruelty to which so many children are prone. But for their teachers to familiarize them with cruelty as part of their training, is a strange perversion of the moral sense.

" I care not whether the anæsthetics are adequate or whether the dissection is of dead animals — *both are revolting and deeply demoralizing for children.* And the enormity is increased where the animals dissected are the companions of our daily life.

"Auguste Comte, who was a philosopher as well as professor of science, taught us that the domestic brutes we train to our service are in a sense admitted to our humanity. And he would not have the highest moral teachers of the young defile themselves with the dissection even of the dead. He thought this was incompatible with the profoundest sense of reverence for human life.

"I write as a parent and teacher of long standing, who has followed courses of philosophy of many eminent men, and who has practical experience of biological experiments."

MISS FRANCES E. WILLARD, President, and LADY HENRY SOMERSET, Vice-President, of the "World's Woman's Christian Temperance Union," have given the following answers to the questions propounded by the *American Humane Association*, in regard to the experimentation upon living animals in the teaching of physiology in the public schools.

1st. In our judgment it must in the nature of the case blunt rather than cultivate the natural sensibilities of children.

2d. The exact opposite is what we believe it is the duty of the teachers of children to inculcate.

3d. We consider it a distinct damage to any child who witnesses such operations.

4th. It is our earnest belief that in view of all the harm that must result from the teaching of vivisection in the public schools, the total results will be incalculably more valuable if teachers would pursue the method you recommend.

5th. By all means.

ERNEST BELL, M. A., *London, England:*

"I have received your circular telling of certain methods of instruction used in schools, and have duly laid it before the members of the committee of the

Humanitarian League, who request me to say that in your efforts to check the increasingly prevalent methods of teaching physiology through demonstrations upon living animals, you have their unqualified approval and hearty sympathy.

'The man who has the most pity is the best man; is the one most disposed to all social virtues, to nobleness of every sort. He who awakens our compassion makes us better and more virtuous,' said Lessing, the great critic; and we may add that he who deadens our pity makes us lower and less virtuous."

RT. HON. JAMES STANSFELD, M. P., *London*:

"I entirely agree with the views of the American Humane Association as expressed in their circular."

HON. GEORGE W. E. RUSSELL, M. P., *London*:

"I have the deepest dislike and distrust of all experiments on living creatures. To practice such experiments before children and young people is in my judgment to give systematic training in brutality.

"The organized destruction of natural feeling is producing its certain results, and I have little doubt that experiments not distinguishable from human vivisection are now of frequent occurrence in hospitals. I wish you all success in your attempt to crush this brutal wrong."

LESLIE STEPHEN, ESQ., *London, England*:

"I am most strongly of the opinion that children should be encouraged in every way to be kind to animals; few practical lessons in morality can, I should judge, be more useful. The dissection of living animals before children appears to be a very doubtful way of impressing such lessons upon them."

DR. GEORGE EBERS, *Munich, Bavaria:*

" The inquiry to the Humane Asssociation I beg to answer as follows : —

" 1. Vivisection is an aid to science, the practice of which, if pursued for earnest scientific research, should not be hindered. On the other hand, experiments which cause pain and even death to helpless creatures, when made in the presence of children in schools, I consider not only useless, but frivolous and harmful as well.

" The parent who wishes to see its offspring have a loving heart will teach it above everything to abhor all cruelty to animals which may be in its power.

" 2. The child knows its power over dumb creatures only too well. To strengthen this knowledge would be useless and injurious, for the child should be taught to respect all living objects and to remember that they were created to enjoy life. I would add that plants should be included in the above. A child that will pick a beautiful flower from a bush and trample upon it, I think has not a good heart, and I know has been badly brought up. A child that will torture an animal for amusement lacks character.

" 3. The educator who wishes to familiarize the pupil with the sight of blood and the act of dying of animals could with more justice burn and cut the child so as to accustom it to pain, for then the body alone would suffer, and not also the soul. There are things which, to become accustomed to, blunt the finer sensibilities and lower the morals, and to these things belong, foremost, the solemn act of dying, or passing away of living beings. Should the child become so hardened and be able to witness the torture and death of animals, it will, when grown up, and having charge of the fate of human beings, be tyrannical and cruel."   .   .   .

HON. ANDREW D. WHITE, LL. D., *Minister to Russia, late President of Cornell University, N. Y.:*

"While I acknowledge that, under very careful restrictions, vivisection may be allowed to men whose character and eminence in appropriate professions give guarantees that their work will be as humanely done as possible and to the best ends, I am utterly and totally opposed to the loose permission to children and youth, and, indeed, to older persons not within the category above referred to." .   .   .   "In my opinion, experiments involving either the infliction of pain or death upon helpless animals in the presence of children should be discouraged."

WILLIAM DEAN HOWELLS, *New York:*

"Vivisection can only be justified in the cause of Science; and though the children's subjects are etherized and suffer no pain, they lose their "little lives" for the sake of imparting a little learning, as useless a knowledge, as vain as any under the sun. Children are shielded by their innocence from many evils; but

I should think such lessons must tend to make them hard and cruel. The whole notion of such instruction is detestable."

PROF. DANIEL G. BRINTON, M. D., *University of Pennsylvania :*

" I believe that physiology can be taught in no other way so successfully as by demonstration on the living subject, and as you and I learned it as physicians in that way, I think that we can both answer that our "natural sensibilities" were not blunted.

" I certainly think that children and every one ought to be familiarized with the sight of blood, the pangs of disease, and the solemn event of dying. Death and pain should not be concealed ; they are the greatest of all educators ; for they alone teach us the value of life in its highest measure.

" The whole tone of your circular is, in my opinion (which you have done me the honor to ask), contrary to the spirit of true education."

MARTIN KELLOGG, A. M., *President of the University of California :*

" Everything needful can be almost entirely taught by use of illustrations or manikins."

NATHAN GREEN, LL. D., *Chancellor Cumberland University, Tennessee :*

" I am unalterably opposed to the dissection of animals such as cats, dogs, etc., before children. The whole business of vivisection is of questionable propriety, and this practice before children for the purpose of instruction is simply barbarous."

PRESIDENT GEORGE WILLIAMSON SMITH, D.D. LL. D. *Trinity College, Hartford :*

"The killing of animals by and before children of public school age, under the plea of instruction in physiology, I am persuaded is unnecessary."

W. J. HOLLAND, *Chancellor Western University of Pennsylvania :*

"As the head of a university, in which the biological sciences and medicine hold a prominent place, I desire to say that in my judgment there is no necessity whatever of familiarizing *children of school age* with the phenomena of death, or with those vital phenomena which are best illustrated by vivisection ; and I question whether in the case of advanced students, except in special cases, which are of necessity very rare, vivisection should be resorted to."

WILLIAM T. HARRIS, A. M., LL. D., *Commissioner of Education, United States :*

" I am glad to learn of some movement against a practice too widely extended of dissecting animals before the children in the elementary schools. I think it well-nigh useless, as far as teaching children a knowledge of anatomy is concerned, and at the same time very injurious to their moral and æsthetic feelings (especially the latter), even when there is no cruelty involved.

" In the high school or academy I think perhaps physiological lessons may be illustrated by the dissection of animals to some extent, but for elementary schools the practice is strongly objectionable."

President William M. Blackburn, A. M., D. D., *Pierre University, Dakota:*

" In my opinion an exaggerated value has been placed upon the study of physiology in the lower grades of our public schools. . . .

" The health lessons and those on temperance do not seem dependent upon physiology when pupils are not capable of the scientific appreciation of the subject. I have doubts whether physiology in any really scientific form or method, *i. e.,* as a science, has been of any very practical benefit in the public schools below the high school grades."

President J. W. Bissell, D. D., *Upper Iowa University:*

" I am fully in sympathy with your efforts to bring about a reformation in our present methods of teaching by vivisection and dissection."

President Edward D. Eaton, D. D., LL. D., *Beloit College, Beloit, Wis.:*

" I fully agree with the *American Humane Society* as to the needlessness and injurious tendencies of the vivisection and even the dissection of animals by and before children of public school age."

James E. Rhoads, LL. D., *President Bryn Mawr College, Bryn Mawr, Pa.:*

" If by ' children ' is meant persons so young that they cannot be expected to appreciate the serious nature of such experiments, the effect will be to blunt their sensibilities. Such ' children ' should never see such experiments.

" If by 'children' be meant those legally so called,
yet from eighteen to twenty-one years of age, these
may witness such experiments provided the experiments are not simply for class instruction but conducted by competent investigators for serious ends.
No vivisection in any form should be used for such
class instruction as is given in public schools or high
schools." . . .

PRESIDENT W. H. PAYNE, PH. D., LL. D., *University
of Nashville, Tenn. :*

" Personally and on deep conviction, I am opposed
to vivisection as practiced in ordinary schools. It is
a needless sacrifice of animal life and has a direct
tendency to blunt and pervert the finer instincts and
feelings of children."

PRESIDENT A. OWEN, D. D., " *Roger Williams University,*" *Nashville, Tenn. :*

" Whatever dulls the sensations to the suffering of
creatures capable of suffering is in every way harmful
to those qualities which most need cultivation and are
most likely to receive it. Too much knowledge of the
system is hurtful. The body is best served by general
obedience to the laws of health and the cultivation of
noble and worthy sentiments."

PRESIDENT JAMES W. STRONG, D. D., *Carlton College,
Northfield, Minn. :*

" Vivisection is unnecessary and barbarous and
nothing of the kind is allowed in connection with our
institution."

PRESIDENT J. BRADEN, D. D., *Central Tennessee College :*

"There is cruelty enough in our land at present. Life is held at too light a value by the great majority of our people."

R. O. BEARD, M. D., *Professor of Physiology, University of Minnesota :*

"To your questions I would make the following answers in order: I think that such experiments as are referred to are likely to blunt the natural sensibilities of children, since their judgment of utility is not educated sufficiently to act independently of emotion excited by the sight of suffering or death. As these emotions are not susceptible of observation or control, they are likely to be destroyed by such influences. In the teaching of children in public schools of the rudimentary truths of physiology and hygiene, everything necessary can be taught by illustrations, manikins, models, and specimens removed from dead animals. . . .

"I appreciate the conservative character of your circular, the more so since it compares favorably with the extreme utterances of anti-vivisection societies. I believe in the utility and morality of vivisection under suitable restriction in scientific schools, but I believe also that the practice needs regulation. In public schools I think it both undesirable and unnecessary."

Rev. J. Percival, D. D., *Head Master Rugby School, England:*

"I am surprised to hear that the method of instruction by means of experiment on living animals is in any degree tolerated in the United States. Happily we are free from it in England."

Prof. Samuel Hart, *Trinity College, Hartford:*

"I find myself entirely in agreement with the principles and practices which the *American Humane Association* maintains. I hope that its advocacy may have a strong influence on public opinion in a practical way."

Edward N. Packard, *Syracuse, N. Y.:*

"I have very decided opinions about the matter, and am strongly opposed to a practice which seems to be prevailing to some extent in our cities. My attention was called to it in this city, and our Ministerial Association in a private way sent me to learn about the practice here in the High School. The agent of the Society for the Prevention of Cruelty to Animals was asked by the teacher of physiology in the 'High' to let her have live dogs for the purpose of experimentation at school. He refused to do it. Further inquiry led to the fact that there was now no vivisection done, but that it had been done. The Board of Education assured us that it would not be allowed. I learned from the mother of a child in the Auburn High School that her son was made sick by seeing the blood, etc., in the operations at the school, and dreaded the day to come for a repetition. I heard that it had been practised for a good while in some Massachusetts schools."

PROF. HENRY C. ADAMS, PH. D., *University of Michigan:*

"I agree fully so far as I understand it with the position taken by the *American Humane Association.* When students have sufficiently advanced to understand the scientific problems now claiming the attention of the medical fraternity about psychology, it may be well to introduce them to vivisection; so far as school children are concerned, it seems to me that a great wrong is being done the children themselves by this means of education."

PROF. FRANCIS E. ABBOTT, *Cambridge, Mass.:*

"While I am not prepared to condemn all vivisection, when conducted by scientific men for strictly scientific purposes and under such conditions as to insure a minimum of pain, I have no hesitation in condemning it unqualifiedly and severely, when it is carrried on in the presence of children, even under the pretence of instruction. Its tendency must be to brutalize them, and this is not atoned for by any mere increase of their knowledge. Legitimate instruction must be in accordance with morality, and it is immoral to inflict pain needlessly on helpless animals. I deny the need of inflicting it for mere illustration and instruction, and can only with extreme reluctance sanction it for purposes of discovery that shall end in lessening it.

"At a time when the need of teaching natural morality, independent of all positive religions, is coming to be widely seen and felt as essential to the conduct of public schools supported by universal taxation, I consider it little short of a crime to teach children to be cruel, or even obtuse to the sight of suffering. And I sincerely applaud the *American Humane Association* for doing what it can to prevent this crime."

PROF. F. TRACY, *Toronto University:*

"In Canada, we have no such experiments as those spoken of in our public schools."

PROF. A. J. GRANGER, *Newton High School, Mass.:*

"As a teacher I should make my answer emphatic. There can be no reason for such experiments in our public schools. I am glad you are fighting this heresy in modern education."

PROF. JOHN B. CLARK, *Amherst College, Amherst, Mass.:*

"I am entirely opposed to vivisection in any ordinary schools for children."

PROF. ALFONSE N. VAN DAELL, *Institute of Technology, Boston, Mass.:*

"I believe that physiology, properly so called, ought not to find a place in any school below the college grade. The elements of hygiene can and ought to be taught, although under present regulations the study, in my opinion, is begun too early.

"I am not opposed to necessary experimentation in colleges, or schools of the same grade, but in public schools experimentation upon living animals is unnecessary or useless, with the possible exception of the senior classes, where something of the real purpose of natural science and comparative anatomy may, under certain circumstances, be admissible."

GEORGE A. BACON, *of Allyn & Bacon, Publishers, Boston, Mass. :*

" To my thinking there is absolutely no excuse for killing animals in order to teach anatomy or physiology in our schools. In the first place the practice in dissection which pupils get amounts to nothing, and they are just as likely to come to wrong as to right conclusions from their observation.

" There is certainly a distinct demoralizing effect produced by familiarity with these details. Any pupil can get from the butcher's shop a sheep's heart and lungs or brain, or sample of bone, muscle or other tissue. All these things lend interest to the subject; they have no appreciable bad effect. The whole object of teaching these subjects in school, or anywhere outside of a medical college, should be simply hygiene. Anatomy and physiology should be made subordinate; adjuncts and handmaids to hygiene.

" I must confess, however, that I do not expect my protest or yours to have very much effect. The cry of the age is all for research, laboratory practice, and that sort of thing. Nothing is supposed to be of any value if learnt in the old-fashioned way. The amusing part of the thing, however, lies in the fact that investigators, unless very skilful, are far more apt to get wrong ideas from direct investigation, than from books."

PROF. H. H. FREER, *Cornell College, Mt. Vernon, Iowa:*

"It is time to call a halt upon the infliction of pain on animals or wantonly killing them for the purpose of teaching anatomy, physiology or hygiene to young children. All that children need to know on these subjects can be taught without resorting to processes that will blunt the sensibilities, deprave the taste and brutalize the whole nature of children.

The boy murderer, Pomeroy, was, I believe, from early life accustomed to the scenes of the slaughter-house, and his environment no doubt was responsible for his cruel and murderous tendencies.

"Your agitation does not involve the general question of vivisection, and should receive the support of all humane persons."

PROF. RAY GREENE HULING, *Head Master English High School, Cambridge, Mass.:*

"Experiments of the sort you describe may tend to blunt the sensibilities of children if performed in their presence.

"Your questions have been carefully phrased and chosen. They still leave opportunity for me to say that I should not object to the use of oysters, sea anemones and similar material in the study of biology by pupils and teachers. . . . In my present school, with the facilities for comparative study of animals afforded by the Agassiz Museum, I prefer such comparative study to a detailed examination of internal structure of familiar animals. Human physiology is illustrated by parts of the pig, the sheep and the ox, regularly, as also by the manikin, the skeleton and pictures. Dissection is not practised."

ALBERT M. HILLIKER, *Washington, D. C.:*

"In answer to the questions of your circular letter will say that I think such experiments as you refer to must necessarily blunt the sensibilities of children witnessing them.

"It is very doubtful whether vivisection can be justified on any ground as practised by any one under any circumstances, and I feel sure that if ever practised it should be by and in the presence of specialists only."

PROF. F. B. KNAPP, *Duxbury, Mass.:*

"I believe in having boys who are especially interested in natural history dissect animals already dead. I do not believe in vivisection before even medical students, but I suppose it is wise for scientists to resort to it to a limited extent."

JOHN E. KIMBALL, *late Superintendent Schools, Newton, Mass.:*

"The practice referred to is unnecessary, painful in the extreme to sensitive natures, cruel and demoralizing. In my experience as Superintendent of Schools I have heard of instances of fainting and real suffer-

4

ing to susceptible children in connection with this very reprehensible practice. If there is one phase of culture outside the usual curriculum in our public schools which should be of constant care, it is the habit of uniform kindness to the lower orders of animate creation, and this is not consistent with a practice which must blunt the sensibilities of all, if it does not in some cases tend to develop types of brutality which from time to time shock society."

Wm. F. Phelps, *late Principal State Normal School, St. Paul, Minn., formerly Principal State Normal School, Trenton, New Jersey:*

"As an educator I would not allow such cruelties to be practised. Experiments upon living animals should be forbidden by statute."

Prof. W. N. Ferris, *Principal Ferris Industrial School, Michigan:*

"I am in sympathy with the work of your Association and do all in my power to advance its interests. I enroll about a thousand pupils every year, two-thirds teachers. I try to impress upon them the principles involved in your Association."

Wm. J. Cox, *Superintendent of Schools, Hancock, Mich.:*

"I am heartily in favor of the good work your Society is doing."

Edward S. Breck, Ph. D., *Boston, Mass.:*

"Although I am in favor of vivisection under certain very rigid restrictions (such as limitation to a few recognized scientific institutions), I am very much against that or anything like it for young people."

. . . . "I think dissection very useful and instructive, but it should be confined to the highest, or the two highest classes in the high schools; and even here its moral effect on the pupils should be carefully noted by the instructors and reported on."

H. D. LLOYD, *Editor "Chicago Tribune:"*

"Experiments involving infliction of pain or death tend to blunt, and therefore to *brutalize*, children in their human relations.

"I do not live up to the doctrine, but I believe that our physical as well as sympathetic evolution is moving to the point at which we will be as incapable of killing animals for food for the body as for food the the mind."

JAMES JEFFREY ROACH, *Editor "Pilot," Boston, Mass.:*

"I consider the vivisection of animals for the ostensible instruction of children to be cruel, useless and demoralizing in the extreme, and that everything necessary for the teaching of physiology could be as clearly and more humanely taught by the use of illustrations and manikins. . . . It is not vitally important that children should know all about their own internal organs; it is absolutely important that they should be taught mercy, even to the lowest of living things."

A. E. DUNNING, *Editor "Congregationalist:"*

"The representatives of the *Congregationalist* do not think vivisection is wise or humane when conducted before classes of boys and girls in the schools. Indeed, the matter seems to me put forcibly and truthfully in the statement with which your circular closes."

J. W. WARR, *Editor "Western Ploughman:"*

"The killing of animals before children is a barbarous practice that ought not to be tolerated in the advanced educational institutions of the nineteenth century."

REV. SAMUEL J. BARROWS, *Editor "Christian Register," Boston, Mass.:*

"I believe it to be a serious mistake to encourage children to any irresponsible use of their power over the lower forms of life.

"Children should be taught that might is not right, and that the same laws of love, mercy and justice, which apply to human beings should be applied to the animal creation as far as possible.

"It seems to me that it is an abuse of the name of education to familiarize children with the infliction upon animals of mortal wounds, etc., under the pretence of imparting scientific knowledge. An animal is not to be treated as a toy which a child is encouraged to take apart just to see how it is put together.

"The development of the spirit of love, mercy and justice is more important than to turn the schoolroom into a butcher's shop or a dissecting-room, to gratify an intellectual curiosity.

"Physiology should have its place in school instruction, but quite as important is the subject of ethics, which includes not only our duties to our fellow-beings, but also our duties to animals."

FINLEY ELLINGWOOD, M. D., *Editor "Medical Times,"* *Chicago, Ill.* :

"I am greatly in favor of physiology being taught children, but I can see no excuse whatever for adopting a course advanced enough to illustrate by vivisection. In the opinion of the Association I coöperate most heartily."

S. T. PICKARD, *Editor "Portland Transcript," Portland, Me.* :

"I am most decidedly opposed to vivisection and dissection before children of public school age. Many grown people would be much happier if they knew less about the possibility of disorder in the organs wisely put out of sight, in order, perhaps, that they might be out of mind."

DR. H. W. PIERSON, *Editor "Medical Advance," Chicago* :

"Promiscuous vivisection is uncalled for and serves to gratify the baser elements in our nature, whether it be children or adults, and should be condemned by all. Individuals preparing for the special study of the subject of physiology will not have their finer senses blunted by study of the mechanism of the body in life. To all others this should be denied, by law if necessary.

"Under sixteen years of age it is not wise to make children familiar with suffering of any kind. Charts and maps are better for the general teaching of the rudiments than the living subject, until the pupil is advanced beyond the elementary exigencies of the study."

RICHARD HOWELL, *Editor "Bridgeport Herald," Bridgeport, Conn.:*

"There are those upon whom vivisection will have a horrifying effect, but there are many in whom the practice in public schools will develop an inordinate love to be cruel to dumb animals.

"The plastic mind of the public school pupil is as sensitive to an impression as the dry plate of a photographer's outfit; and the impression which vivisection makes upon one of these young minds may develop frightful traits of character."

ERNEST H. MORGAN, *Editor " Roxbury Gazette," Roxbury, Mass.:*

"I am against vivisection, even among advanced students, and utterly and uncompromisingly opposed to it among pupils in public schools."

J. SILVERSMITH, *Editor " Occident," Chicago, Ill.:*

"I believe the rudiments of physiology and hygiene can be taught very well without resort to vivisection."

REV. E. B. GRAHAM, *Editor "Midland," Chicago, Ill.:*

"Children should not be allowed to see game shot by cruel sportsmen, or domestic fowls killed even for food, and much less should they become familiar with cruelty in the interests of education."

MILTON E. SMITH, *Editor " Church News," Washington, D. C.:*

"I fully sympathize with any movement which tends to make children realize that it is ungentlemanly, inhuman and contrary to the spirit of civilization to inflict unnecessary suffering upon either man or brute."

DR. M. L. HOLBROOK, *Editor "Herald of Health:"*

" I do not think the slightest good in practice ever comes to children from the experiments alluded to. They are unnecessary. Study animals alive, acting naturally, and some good can be learned. Studying them in the throes of pain cannot help teach hygiene."

WM. NORTON PAYNE, *Editor "Dial," Chicago, Ill.:*

" In my opinion dissection has a necessary place in the school work, but vivisection of vertebrates should not be tolerated."

J. W. BASHFORD, *Mt. Vernon St., Boston, Mass.:*

" I believe the older children in our public schools would be benefited by actual knowledge of the structure of animals, and would gain thereby greater reverence for all life. But I think in general that it would be wise that demonstrations be upon animals used for food."

CHARLES W. STONE, *Boston, Mass.:*

"I wish I had time to set forth at length my utter detestation of this outrageous perpetration in the name of education."

MISS HELEN C. HAWKINS, *Tolland, Conn.:*

" My experience as a teacher has convinced me that boys are apt to treat animals ungently and even cruelly. In most cases the thought of suffering on the part of the animal has never presented itself until it has been presented by those who have most to do with their early training."

MRS. EDNAH D. CHENEY, *Boston, Mass.:*

"I think it hardly wise to introduce much special physiological instruction into schools of the lower grades, unless under the care of most judicious teachers, which we can hardly expect all to be. In general, I should object to experimenting with living subjects, as of little use to such young pupils and liable to great abuse. Observations regarding the life and habits of animals I think more valuable, and this can be much encouraged."

T. A. ABBOTT, ESQ., *St. Paul, Minn.:*

"Our St. Paul schools, although having a department of physiological science of the highest excellence, are opposed in theory and practice to vivisection."

FRED P. BAGLEY, ESQ., *Chicago, Ill.:*

"The truths of physiology can be taught as well by the use of illustrations and manikins as by dissection; there is no necessity to resort to experiments upon living creatures."

HON. AUSTIN V. EASTMAN, *St. Paul, Minn.:*

"I most heartily agree with the suggestions contained in the circular, and am strenuously opposed to conducting experiments in public schools in the manner outined therein."

CHARLES A. HAMLIN, ESQ., *Syracuse, N. Y.:*

"All experience proves that familiarity with cruelty,

pain, and suffering renders men increasingly indifferent to it, and withers the sense of pity."

MISS ALICE M. LONGFELLOW, *Cambridge, Mass.:*

"It would seem to be of far greater value to lay stress upon the importance of observing and understanding a living creature, instead of taking away the essential element of its beauty and interest. It seems to be poor humanity and poor science to think either is served by destruction instead of by preservation."

CLIFFORD W. BARNES, *Chicago, Ill.:*

"Having studied physiology and hygiene by the use of illustrations and manikins, and having afterwards studied in a medical college and had experiments in vivisection, I can speak with assurance when I say that no child in the public schools needs to resort to experimentation on living creatures in order to obtain a perfectly satisfactory and sufficient knowledge of the essentials of physiology."

PROF. JOHN TROWBRIDGE, S. D., *Harvard University, Cambridge:*

"I have no hesitation in saying that I agree entirely with the position that teaching physiology by vivisection in public schools, is brutalizing and unnecessary."

In response to the circular, replies were received from a large number of persons, for the most part expressing sympathy and accordance with the position of the *American Humane Association* on the question of vivisection in schools, but of whose valued letters considerations of space prevent more than a brief acknowledgment. In many cases, too, the circular was answered simply by monosyllables or marginal notes. To the following persons, therefore, the thanks of the *American Humane Association* are also due for responses to its circular:

Prof. DORMAN B. EATON, *New York.*

E. L. GODKIN, ESQ., *Editor of "New York Evening Post."*

ALFRED H. LOVE, *President American Peace Society, Philadelphia, Pa.*

FLOYD W. TOMKINS, Jr., *Rector Grace Church, Providence, R. I.*

REV. WILLIAM BRUNTON, *Whitman, Mass.*

REV. DR. HENRY BLANCHARD, *Portland, Maine.*

REV. DR. E. M. HICKOK, *Sharon, Mass.*

REV. L. WEISS, *Columbus, Ohio.*

REV. HENRY COHEN, *Galveston, Texas.*

REV. ENDICOTT PEABODY, A. M., *Groton School, Groton, Mass.*

CAROLINE T. HAVEN, *Workingman's School, New York City.*

J. VAN INWAGEN, ESQ., *Chicago, Ill.*

G. E. MORROW, *President Agricultural Experiment Station, University of Illinois.*

RT. REV. GEORGE D. GILLESPIE, *Bishop of Western Michigan.*

RT. REV. HUGH MILLER THOMPSON, *Bishop of Mississippi.*

Rt. Rev. Jos. Blount Cheshire, Jr., *Bishop of North Carolina.*

Rt. Rev. Henry B. Whipple, *Bishop of Minnesota.*

Rt. Rev. M. A. DeWolfe Howe, LL. D., *Bishop of Pennsylvania.*

Rt. Rev. Thomas Bowman, *Bishop M. E. Church, St. Louis, Mo.*

Rev. Dr. A. S. Fiske, *Ithaca, N. Y.*

Rev. Dr. D. F. Bonner, *Pastor Presbyterian Church, Florida, N. Y.*

Rev. Edward C. Hood, *Wrentham, Mass.*

Rev. Austin S. Garver, *Worcester, Mass.*

Rev. Dr. Egbert C. Smyth, *Andover, Mass.*

Rev. Dr. C. H. Eaton, *New York City.*

Rev. Paul Van Dyke, *Northampton, Mass.*

Prof. William Knight, *University of St. Andrew's, Scotland.*

Rabbi Max Wertheirmer, *Dayton, O.*

Rev. William R. Campbell, *Roxbury, Mass.*

Rev. Dr. George W. Wood, *Mt. Morris, N. Y.*

Rev. Dr. I. J. Lansing, *Park Street Church, Boston, Mass.*

Rev. Dr. William R. Campbell, *Roxbury, Mass.*

Rev. Francis M. Collier, *Denver, Col.*

Rev. Daniel L. Furber, *Newton Centre, Mass.*

Rev. Dewitt M. Benham, *Pittsburgh, Pa.*

Rev. E. C. Ewing, *Danvers, Mass.*

Rev. George Sexton, D. D., M. D., *Dunkirk, N. Y.*

Rev. Dr. Joseph H. Jenckes, *Indianapolis, Ind.*

Rev. Stephen Peebles, *Satank, Col.*

Rev. Wm. E. Barton, *Shawmut Church, Boston, Mass.*

Rev. Dr. Edward Abbott, *Cambridge, Mass.*
Rev. Alsop Leffingwell, *Philadelphia, Pa.*
Rev. Dr. Alex. G. Wilson, *Pres. Theological Seminary, Omaha.*
Rev. Austin B. Bassett, *Ware, Mass.*
Rev. Dr. James H. Potts, *Editor "Michigan Christian Advocate," Detroit, Mich.*
Rev. Paul P. Frothingham, *New Bedford, Mass.*
Rev. Dr. Reese F. Alsop, *Brooklyn, N. Y.*
Rev. James B. Gregg, *Colorado Springs, Colo.*
Rev. C. H. Rogers, *Oklahoma.*
Rev. J. Vila Blake, *Chicago, Ill.*
Rev. T. G. Ensign, *Superintendent of the American Sunday School Union.*
Rev. James H. Darlington, *Christ Church, Bedford Ave., Brooklyn, N. Y.*
Rev. John P. Coyle, *North Adams, Mass.*
Rev. J. M. Williams, *Burlington College, Burlington, N. J.*
Rev. Wm. Cleveland Hicks, Jr., *New York City.*
Rev. A. W. Meyer, *Editor "Lutheran Guide."*
Rev. Dr. Epher Whitaker, *Southhold, N. Y.*
Rev. Dr. James Roberts, *Colwyn, Pa.*
Rev. Charles W. Wendte, *Oakland, Cal.*
Rev. A. W. Jackson, *Concord, Mass.*
Rev. Dewitt S. Clark, *Chairman of High School Committee, Salem, Mass.*
Rev. Dr. George F. Kenngott, *Lowell, Mass.*
Rev. William Lloyd Himes, *Concord, N. H.*
Rev. Charles H. Oliphant, *Methuen, Mass.*

Rev. T. H. M. Villiers Appleby, M. A., *Archdeacon of Minnesota.*

Rev. Dr. Charles J. Jones, *Stapleton, N. Y.*

Rev. Thomas Duck, *Hammondsport, N. Y.*

Prof. William Knight, *University of St. Andrews, Scotland.*

President George A. Gates, D. D., *Iowa College, Grinnell, Ia.*

President W. H. Wilder, D. D., *Illinois Wesleyan University, Bloomington, Ill.*

President Franklin Carter, Ph. D., LL. D., *Williams College, Mass.*

President William Preston Johnston, LL. D., *Tulare University, New Orleans.*

President Julius D. Dreher, A. M., Ph. D., *Roanoke College, Va.*

President William G. Frost, Ph. D., *Berea College, Kentucky.*

President J. E. Rankin, D. D., LL. D., *Howard University, Washington, D. C.*

President William A. Obenchain, A. M., *Ogden College, Bowling Green, Ky.*

President J. B. Shearer, D. D., LL. D., *Davidson College, Davidson, N. C.*

Governor L. Bradford Prince, LL. D., *President University of New Mexico.*

President Daniel A. Long, D. D., LL. D., *Antioch College, Ohio.*

President John V. N. Standish, Ph. D., *Lombard University, Galesburg, Ill.*

W. Scott Thomas, *Superintendent Public Schools, San Bernardino, Cal.*

Francis Coggswell, *Superintendent Schools, Cambridge, Mass.*

W. B. Powell, *Superintendent of Public Schools, Washington, D. C.*

P. W. Search, *Superintendent Schools, Pueblo, Col.*

Miss R. S. Rice, A. M., *Principal Girls' Collegiate School, Chicago, Ill.*

Miss Sara J. Smith, *Principal Woodside Seminary, Hartford, Conn.*

Prof. W. C. Sawyer, Ph. D., *University of the Pacific, California.*

Prof. H. M. Willard, *Howard Seminary, Mass.*

Prof. Edward A. Allen, *University of Missouri.*

Rev. R. W. Chestnut, *Editor " Reformed Presbyterian Advocate," Marissa, Ill.*

E. C. Linfield, *Editor "Duxbury Breeze."*

George M. Whitaker, *Editor " N. E. Farmer," Boston, Mass.*

E. H. Clement, *Editor " Boston Transcript."*

James P. Magenis, *Editor "Adams Freeman," Adams, Mass.*

Rev. Wm. Dallmann, *Editor " Lutheran Witness," Baltimore, Md.*

R. H. Carothers, *Editor " Educational Courant," Louisville, Ky.*

J. M. Dewbery, *Editor " Educational Exchange," Montgomery, Ala.*

W. J. Chalmers, *Chicago, Ill.*

Maria H. Blanding, *Girls' High School, Brooklyn, N. Y.*

Miss Maria L. Owen, *Ex-President Springfield Women's Club, Springfield, Mass.*

Miss E. E. Constance Jonès, *Girton College, Cambridge, England.*

Miss Rowena A. Pollard, *Georgetown, Ky.*

Grace A. Oliver, *Marblehead, Mass.*

Miss Stella Dyer Loring, *Prairie Ave., Chicago, Ill.*

Dean L. B. R. Briggs, *Harvard University, Cambridge.*

Wm. C. Collor, *Keene Valley, N. Y.*

J. W. Plummer, *Chicago, Ill.*

E. N. L. Walton, *West Newton, Mass.*

Charles C. Pickett, Esq., *Chicago, Ill.*

H. R. Arndt, M. D., *San Diego, Cal.*

George Sadler, M. D., *Ravenna, Ohio.*

Christopher Roberts, Esq., *Newark, N. J.*

L. F. Ives, Esq., *Detroit, Mich.*

Calvin M. Clark, *Haverhill, Mass.*

Marion Lawrence, *General Secretary, Ohio Sunday School Association, Toledo, Ohio.*

Otto Reiner. Esq., *Brooklyn, N. Y.*

C. B. Grant, Esq., *Houghton, Mich.*

T. Griswold Comstock, M. D., *St. Louis, Mo.*

F. Wilson Hurd, M. D., *Munsi, Pennsylvania.*

Ada H. Kepley, *Attorney-at-Law.*

Hon. John Turner Wait, *Norwich, Conn.*

All of which is respectfully submitted.

Francis H. Rowley,

Albert Leffingwell, M. D.,

Committee.

www.ingramcontent.com/pod-product-compliance
Lightning Source LLC
Chambersburg PA
CBHW022008190326
41519CB00010B/1438